Math Activities with
DOMINOES

Helene Silverman
Sandy Oringel

Cuisenaire Company of America, Inc.
White Plains, New York

Managing Editor: Doris Hirschhorn
Development Editor: Harriet Slonim
Design Director: Phyllis Aycock
Cover Design and Illustration: Tracey Munz
Text design, line art, and production: Tracey Munz

4 5 00 99 98

Table of Contents

Introduction 5

Getting Started with Dominoes 8

1. Let's Begin 9

    Get on Board 10
      find the dominoes that comprise a suit, then find a domino shared by two suits

    I'm All Set 12
      sort dominoes by suit to form a Venn diagram

    Parking Lot 14
      match zero-suit dominoes to number cards that correspond to their non-zero faces

    Lots of Spots 16
      match domino faces to spot cards and compare values of the faces

    Domino Dash 18
      match the pips on one face of a zero-suit domino to a roll of a number cube

    Follow Suit 20
      sort dominoes by suit and make a graph

2. Relationships 21

    Turn Arounds 22
      identify a domino in two different positions

    Cross-ups 24
      find the double dominoes that can bound a single domino

    Surround 26
      identify the suit associated with a double domino

    Find My Neighbors 28
      identify dominoes whose values are greater than or less than the value of a given domino

    Follow the Leader 30
      compare the values of several dominoes to a single domino

    Check It Out 32
      compare the values of three dominoes to each other

3. Sort and Classify 33

    Let's Graph! 34
      sort dominoes to form a graph

    This Way or That? 36
      sort and classify dominoes by their attributes

    Math Marvel 38
      sort dominoes according to their value and solve a simple probability problem

    Who Am I? 40
      identify a domino described in a riddle

    Double-Circle Sets 42
      sort dominoes according to their attributes to form a Venn diagram

4. What's It Worth? 45

Is It a Tie? 46
play a game to compare the values of two or more dominoes

Play for Ten 47
play games combining dominoes in different ways to form sums of 10

Subtraction Action 48
play a game to find the difference between the values of two faces of a domino

Diffy 50
play a game to find differences between values of domino pairs

Pip-Pip-Hooray! 52
play a bingo-like game to find the values of dominoes

5. Think It Through 53

Number Neighbors 54
combine specific dominoes to form target sums

Match Mates 56
play a memory game to identify pairs of dominoes with equal sums

Circle, Circle Sums 58
combine dominoes to form equal sums and represent them in a Venn diagram

Memory Mates 60
play a memory game to identify pairs of dominoes with equal sums or differences

Target 20 61
find different ways to reach a target number

Lucky Roll 62
find the sum of the values of domino faces or the difference between them

6. Connecting Paths 63

Path of Nine 64
match the faces of nine dominoes to form a continuous path

Hidden Path 66
use logic to find dominoes to complete a path

Pattern Paths 68
use dominoes to identify and continue patterns

Along the Path 70
play a game to match domino faces

Take Five 71
match dominoes to create a path longer than other players' paths

Triangle Paths 72
match domino faces to form triangular paths

Square Paths 74
match domino faces to form square paths

Equal Paths: Triangles & Squares 76
complete paths each of whose sides have the same value

R.A.P. (Roll, Add, Place) 77
match a domino face with the missing addend to reach a sum

Blackline Masters 78–80

Introduction

Math Activities with Dominoes provides hands-on, child-centered math tasks for children in kindergarten through grade three. The activities encourage children to investigate basic math concepts, construct their own mathematical meanings, and communicate their ideas. As they explore with dominoes, children have the opportunity to become confident problem solvers, developing mathematical thinking abilities, and assimilating concepts. This is consistent with the vision proposed by the National Council of Teachers of Mathematics (NCTM) in the *Curriculum and Evaluation Standards for School Mathematics*. The first four standards called for involve problem solving, communication, reasoning, and making mathematical connections.

About Dominoes

Dominoes are available in double-six and double-nine sets. Each domino piece is divided into two squares, called *faces*. Each face has a value determined by a number of spots, called *pips*.

The double-six set of dominoes consists of 28 pieces. Each piece has from 0 to 6 pips on each of its faces. Seven of the pieces, called *doubles,* have equal numbers of pips on their faces.

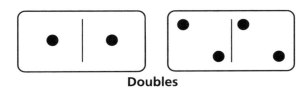

Doubles

The remaining 21 dominoes, called *singles,* have two different numbers of pips on each face. A domino face with no pips is called a *blank,* or *zero.*

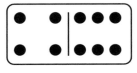 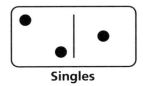 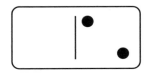

Singles

The double-six piece is the domino with the greatest value. The double-blank, or double-zero piece, is the domino with the least value.

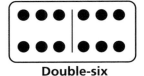 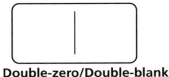

Double-six   **Double-zero/Double-blank**

Math Activities with **DOMINOES**

The values 0 to 6 are represented in each domino suit. There are seven suits, each with seven dominoes. Except for the doubles, each domino is a member of two different suits. For example, the zero-suit contains the following dominoes:

Zero-suit dominoes

The three-suit contains these dominoes:

Three-suit dominoes

Notice that the 0–3 domino, in the zero-suit is the same as the 3–0 domino in the three-suit, and so it is in both suits.

The complete 28-piece set of double-six dominoes have pips on their faces representing the suits from 0 to 6:

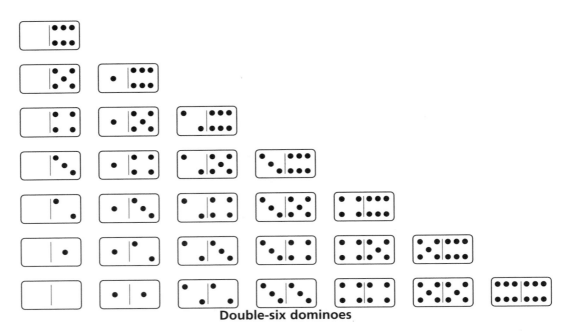
Double-six dominoes

The complete 55-piece set of double-nine dominoes have pips on their faces representing the suits from 0 to 9. The seven-, eight-, and nine-suit dominoes are added to the double-six set to make up the double-nine set.

About the Activities

In preparation for working with the activities, multiple sets of dominoes should be acquired so that children can work in pairs or small groups. The activities can be modeled for the class on the chalkboard or on an overhead projector. They can be modeled for a small group of children on a desk or table. After an activity is explained and/or modeled for the class, children can work on the activity in pairs or small groups, either at their desks or at a math center. Most of the activities are explained for use with a double-six domino set, but they can be easily modified for use with a double-nine set of dominoes.

Mats accompanying most activities serve as work areas or game boards. Many of the activities are games which children will want to play over and over again. The mats may be duplicated, colored with markers, and/or laminated for repeated use. Copies of mats may be duplicated for children to take home so that they can share their work and play the games with family members.

The activities are organized into six chapters, each emphasizing a major mathematical concept for each activity. The teachers' notes for each activity are organized as follows to help the teacher and children get the most from the activities:

Task gives the main objective, skills developed, or purpose.

Set-up indicates the materials needed and suggested groupings of children.

Start-up describes the activity, procedure, or game rules.

Discussion focuses on how children generally respond to the task and ways to address questions which may arise. Sample solutions are often provided.

Keep-up provides ideas for keeping the task on target.

Wrap-up suggests critical thinking questions for class discussion or for use in making journal entries.

Follow-up expands the task for the more experienced students.

About the Activity Features

- Each activity encourages children to solve problems through the use of process skills such as describing, sorting, matching, ordering, listing, counting, comparing, and patterning. Children are often asked to think about how they solved a problem so that they become confident in their own reasoning ability. Often, there are different solutions to a problem and different strategies that can be used to solve it.
- The emphasis of the activities is on investigation. Children are encouraged to record and share their ideas.
- The activities are designed to be open-ended and can easily be adapted to meet the needs of any class. Game rules may be altered to simplify an activity or to make it more challenging.
- The activity mats are designed to be reproduced and distributed.
- Blackline masters, found on pages 78–80, can be duplicated for use with many of the activities.

Getting Started with Dominoes

Task   Children are introduced to the characteristics of a set of dominoes.

Set-up   A small group of children uses one set of dominoes.

Start-up   Display a few dominoes and ask children to share what they know about dominoes. As children point out the features, encourage a variety of responses. Hold up two dominoes. Ask children to tell how the faces of these dominoes are alike and/or how they are different. Do this several times with different dominoes.

Discussion   The domino terminology explained in the "About Dominoes" section of the Introduction should become an integral part of children's vocabulary as they work with dominoes. Children should know that in a double-six set there are 28 dominoes and each one is different from the others. They should be able to identify the two faces of each domino according to their number of pips. When comparing several dominoes, children may find patterns. Encourage them to point these out to the rest of the group.

Keep-up   Children can choose any two dominoes from the set, then talk with their partners about the likenesses and differences between the faces. You may want to have some children record the likenesses and differences.

Wrap-up   Key questions for discussion or response in journals:

   • What are some ways in which two dominoes can be alike?
   • What are some ways in which two dominoes can be different?

Follow-up   Use the double-nine set of dominoes to extend the activity. Have children choose three dominoes from this set and tell about their likenesses and differences. Children may want to choose two or three particular dominoes in order to show specific likenesses and differences rather than randomly selecting dominoes.

Let's Begin

Children use observation, classification, matching, and counting as they engage in activities that introduce them to the language and patterns of dominoes. They develop number sense and pattern recognition as they match dominoes to activity mats. Spot cards, number cards, dice, and number cubes provide multiple forms of reinforcement of numerals and pip patterns.

Task   Children find the dominoes that comprise a complete suit, from 0 to 6. They compare two suits and identify the domino that belongs to both.

Set-up   Each pair of children uses one set of dominoes and a set of 0–6 number cards (page 79). Each child has a "Get on Board" mat.

Start-up   Tell children they will each make a number train. Have them each select a number card for the engine of their train. Explain that the number card names a domino suit. Have children find the dominoes in that suit and put them on the train. Partners should then compare their trains and discuss what they notice.

Discussion When children have identified the dominoes that belong on each train, they will notice that the dominoes all have one face that identifies the suit and one each of the faces 0 to 6. They will notice that they cannot show two full suits at once since any two suits share one domino.

Keep-up   Challenge pairs to position the dominoes of two suits showing how they share one common domino.

One sample solution:

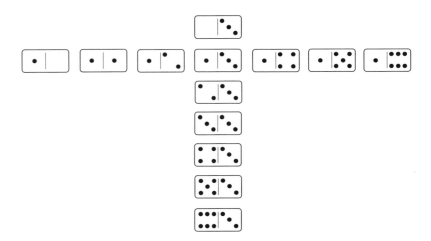

Wrap-up   Key questions for discussion or response in journals:
- How can you tell if you have all of the dominoes in a suit?
- Did you have a problem showing all the dominoes in two suits at the same time? If so, describe the problem.

Get on Board

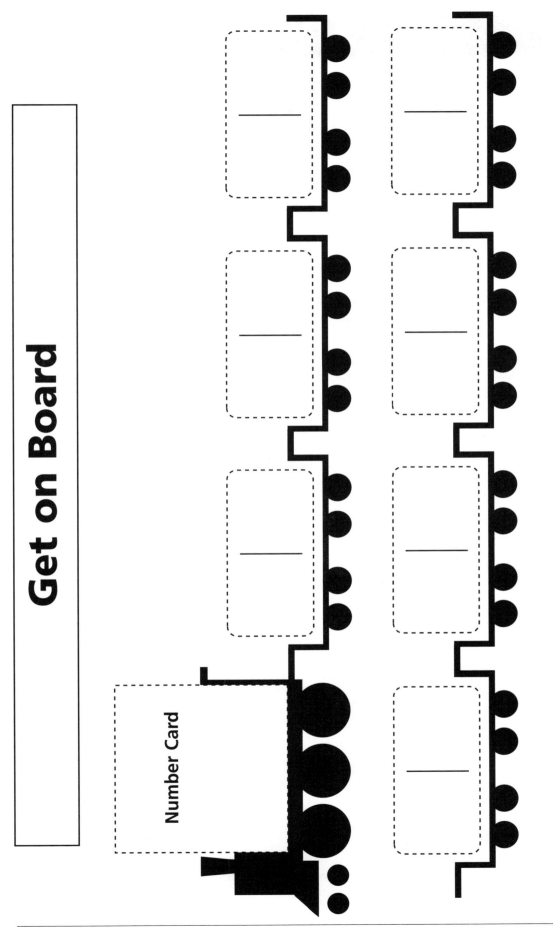

Number Card

Math Activities with **DOMINOES** **11**
©1996 Cuisenaire Company of America, Inc.

I'm All Set

Task   Children sort dominoes by suits to form a Venn diagram.

Set-up   Each pair of children uses one set of dominoes, a set of 0–6 number cards (page 79), and an "I'm All Set" mat.

Start-up   Each child draws one number card and places it in one of the boxes on the "I'm All Set" mat. Children sort the dominoes on the mat according to the suits named by the number cards. They repeat this procedure for another two suits. The dominoes of one suit should be put in the circle on the left. The dominoes of the other suit should be put in the circle on the right. The domino common to both suits should be put in the overlapping part, or intersection, of the circles.

Discussion   Have children explain why one domino belongs in the intersection of the circles.

One sample solution:

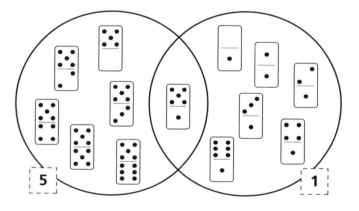

Wrap-up   Key question for discussion or response in journals:
                  • How can you tell which domino is shared by both suits?

Let's Begin

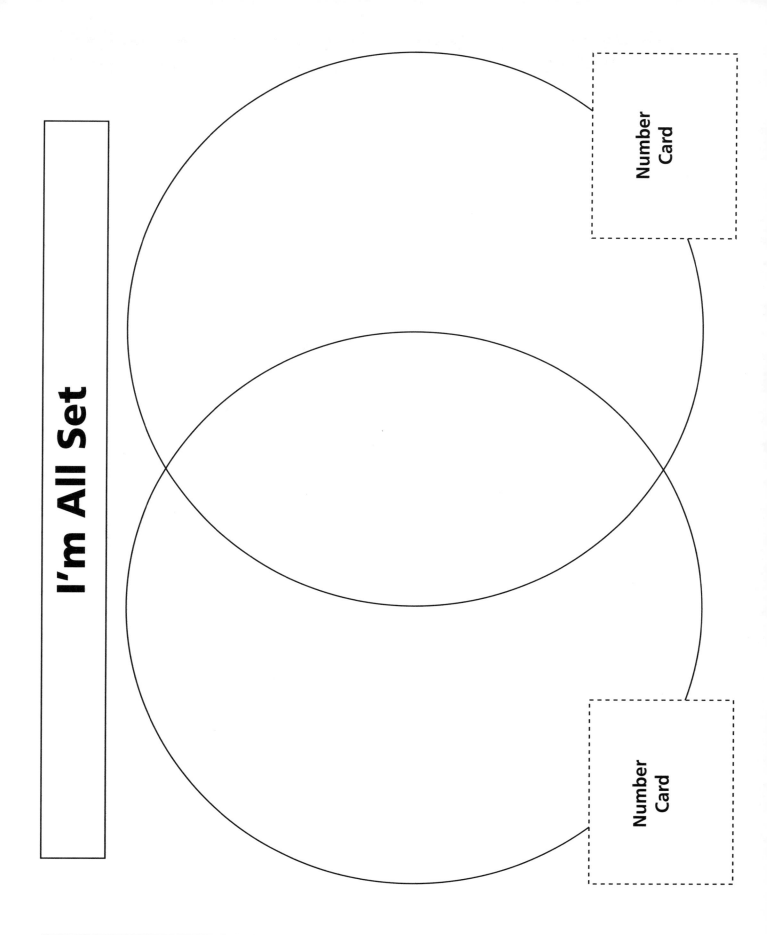

I'm All Set

Number Card

Number Card

Math Activities with **DOMINOES** **13**

©1996 Cuisenaire Company of America, Inc.

Parking Lot

Task Children match six of the zero-suit dominoes to the number cards that correspond to their non-zero faces.

Set-up Each pair of children uses a set of dominoes, a set of 1–6 number cards (use sum cards on page 80), and a "Parking Lot" mat.

Start-up Have children group together the zero-suit dominoes. Tell them to remove the double-zero domino. Then have children put one number card in each space on the "Parking Lot" mat. Have them then place the dominoes on the mat in the corresponding parking spaces, then check to be sure that the matches are correct.

Discussion After children have positioned their cards and dominoes, they should order them according to their values, from least to greatest.

Wrap-up Key question for discussion or response in journals:
- How can you be sure that each domino is in the correct parking space?

Follow-up Challenge children to arrange the number cards in a different way, then move the dominoes accordingly.

Parking Lot

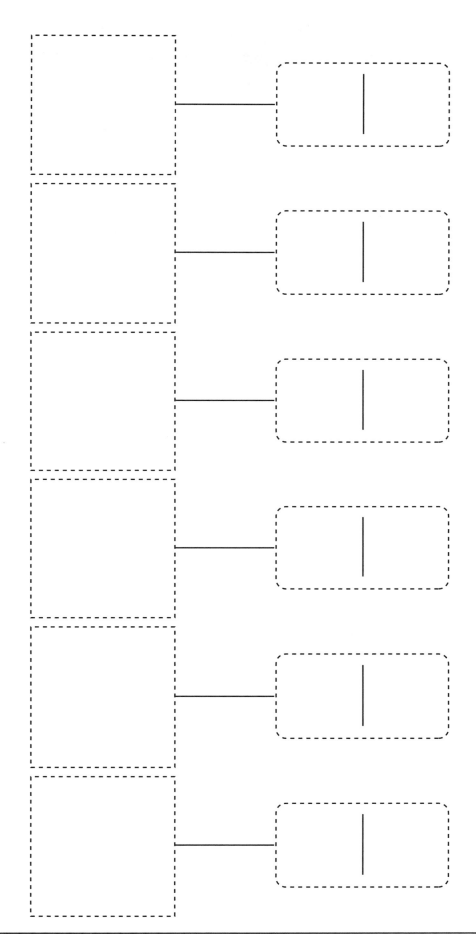

Park number cards here.

Park dominoes here.

Lots of Spots

Task
Children match the two faces of a domino to two "spot cards." They determine which face has more pips, fewer pips, or whether they have the same number of pips.

Set-up
Each pair of children uses a set of double-six dominoes and two sets of 0–6 spot cards (page 78).

Start-up
One child chooses a domino without looking and places it at the top of the mat. Each child finds the spot card that matches one of the faces of that domino. The children compare their cards and say which face has more pips or fewer pips or whether they have the same number of pips.

Discussion
You may wish to model the process with children before they work in pairs. Ask them how they know which face of the domino has more pips or fewer pips or whether the two faces have the same number of pips. If children do not know how to compare, they should together count their pips until one child "runs out of pips." Alternately, they can match counters and pips in one-to-one correspondence.

Keep-up
Challenge children to devise a way to record the ways in which the two faces of their dominoes compared.

Wrap-up
Key questions for discussion or response in journals:
- How did you decide which face had more pips or fewer pips or whether the faces had the same number of pips?
- Which domino in the set has the most pips?
- Which domino in the set has the fewest pips?

Follow-up
Have two children each select a spot card and put it on the bottom half of the mat. Tell them to take turns counting aloud as they place small counters in one-to-one correspondence with the pips on their spot cards. Together children should find the domino that matches their spot cards. They can match counters in one-to-one correspondence to determine which face has more pips.

Lots of Spots

Domino

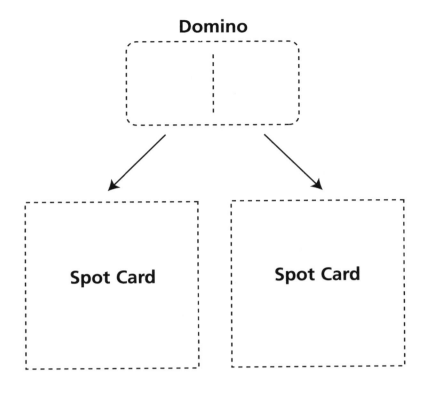

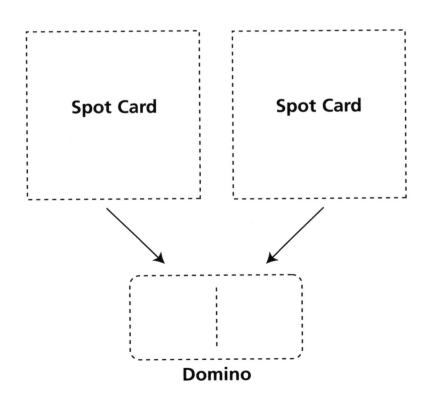

Domino

Domino Dash

Task

Children match the pips on one face of a zero-suit domino to the numeral rolled on a number cube.

Set-up

Each pair of children uses one set of dominoes and a 1–6 number cube. Each child has a "Domino Dash" mat. A green crayon or marker is optional.

Start-up

Children find the zero-suit dominoes and remove the double-zero. Each child chooses three of the zero-suit dominoes and places them on the mat to begin each path. Children take turns rolling the number cube. The child having the domino with a face that matches the number rolled moves that domino one space along the path. If a domino lands on a space marked "GO!" it immediately moves ahead one more space. The first domino to cross the finish line at the end of a path wins. For each game, new dominoes are chosen.

Discussion

When children have picked their dominoes, they should count the pips and identify each one. For children having difficulty recognizing numerals, spot cards could be picked instead of rolling a number cube. It may be helpful to children to remember how to move if they land on "GO!" if they color that word with a green crayon or marker.

Keep-up

To vary the activity, have children keep playing until one player has all three dominoes cross the finish line.

Wrap-up

Key question for discussion or response in journals:

• Do you think one domino will win more often than another? Explain.

Follow-up

Have three children play the game with the double-nine set of dominoes and pick from a set of 1 to 9 number cards (page 79) or roll a pair of dice and use the sum.

Domino Dash

Cross the finish line to win!

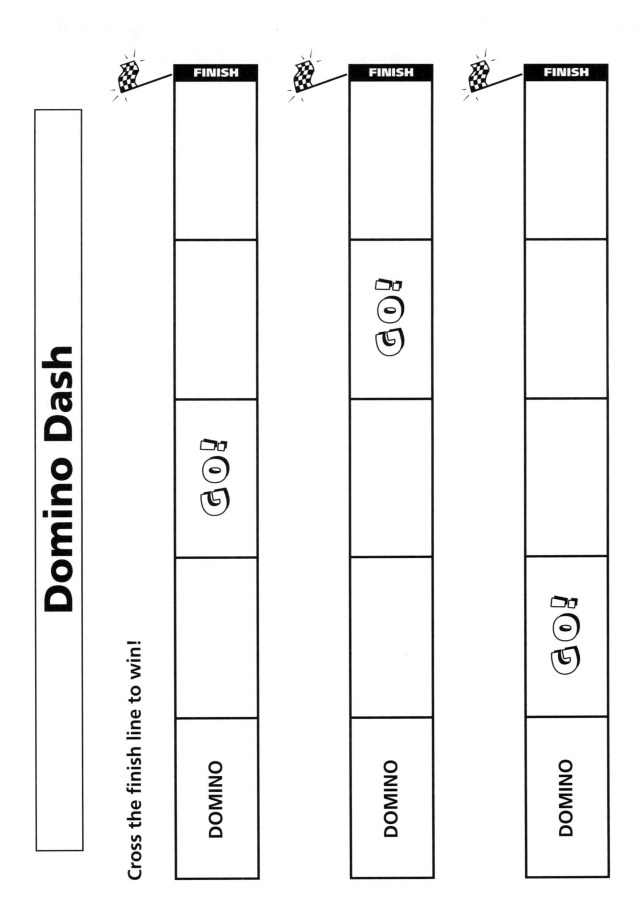

Follow Suit

Task
Children sort dominoes by suits to form a graph.

Set-up
Each pair of children uses one set of dominoes and a set of 0–6 number cards (page 79).

Start-up
Have children predict which suit will have the most dominoes in it when they sort by suit. Tell children to line up the number cards from 0 to 6 and, starting with 0, sort the dominoes by suit, placing them in columns. Once a domino is placed in one column, it should not be moved to another.

Discussion
Have children share their predictions. Most children will make guesses, rather than predictions, unless they have had similar experiences. When the children have sorted the dominoes by suit in ascending order, they will notice that some of the suits are incomplete because some dominoes were used in a previous column. Encourage them to locate the "missing" dominoes. Children should describe their graphs and compare them with others.

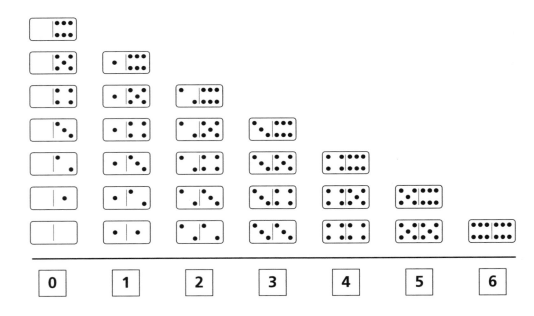

Wrap-up
Key questions for discussion or response in journals:
- How did you know where to place a domino?
- What does your graph look like?

Follow-up
Challenge children to sort and graph their dominoes in another way.

Relationships

Children use observation, matching, ordering, and comparing to explore relationships among the dominoes. They engage in activities in which they use pattern recognition, cardinality, and ordinality to arrange the dominoes in a variety of ways.

Turn Arounds

Task   Children identify a domino in two different positions.

Set-up   Children work independently using one set of dominoes and a "Turn Arounds" mat.

Start-up   Have children each pick a domino and place it on the first dotted outline on the mat. Tell them to copy their domino, then turn it around and copy it again to show how it looks turned around. Emphasize that while turning a domino changes how it looks, it does not change the domino. Repeat this procedure several times.

Discussion Ask children to compare the faces of each domino—before and after it was turned. Have them explain why a single domino can look different. Children will notice that a double will look the same before and after it is turned.

Keep-up   Challenge children to find the dominoes that look exactly the same regardless of how they are turned.

Wrap-up   Key questions for discussion or response in journals:

- How does a domino change when you turn it around?
- Does the number of pips on a domino change when you turn it around? Why?

Turn Arounds

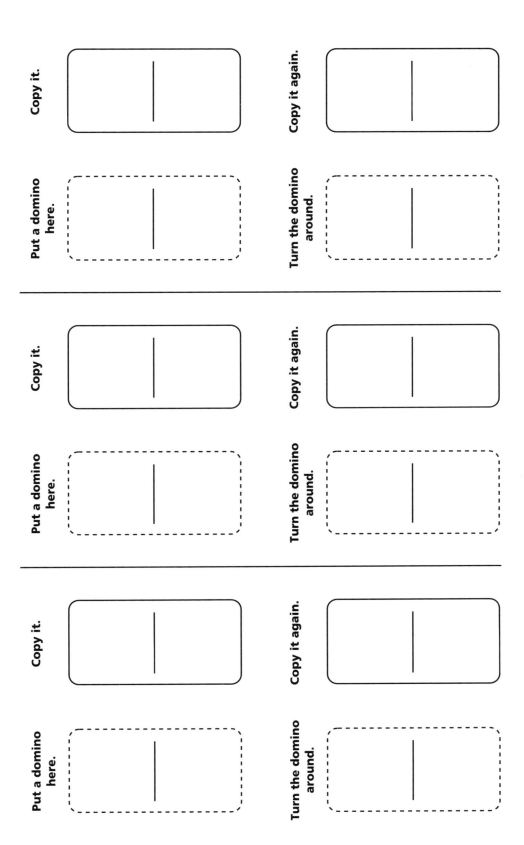

Copy it.

Put a domino here.

Copy it again.

Turn the domino around.

Copy it.

Put a domino here.

Copy it again.

Turn the domino around.

Copy it.

Put a domino here.

Copy it again.

Turn the domino around.

Math Activities with **DOMINOES** **23**
©1996 Cuisenaire Company of America, Inc.

Cross-ups

Task   Children find the two domino doubles that can bound a single domino.

Set-up   Each pair of children uses one set of dominoes and a "Cross-ups" mat.

Start-up   One partner selects any single domino and places it horizontally on an outline on the "Cross-ups" mat. The other partner finds the doubles that match the faces of that domino and places each vertically next to the matching face. Children take turns selecting the single domino and finding the doubles. They repeat the procedure several times.

Discussion You may want to model the activity for some children. Ask children how they know which doubles to find for a given single domino.

One possible solution:

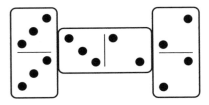

Keep-up   Challenge children to place three single dominoes, along with the doubles that bound them, on the mat. Ask children how they selected their singles.

Wrap-up   Key question for discussion or response in journals:

- What did you have to do to show three single dominoes and their matching doubles at the same time?

Follow-up   Have children pick a domino without looking and place it horizontally on an outline on the "Cross-ups" mat. Have them identify the number of pips on one face. Then have them bound that domino with another, the sum of whose faces equals that number. Have them bound the other face in the same way.

One possible solution:

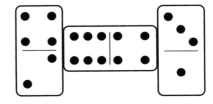

Relationships

Cross-ups

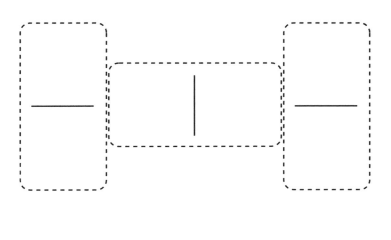

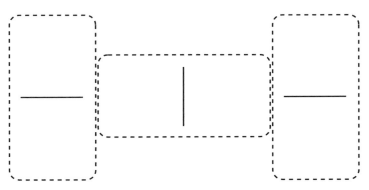

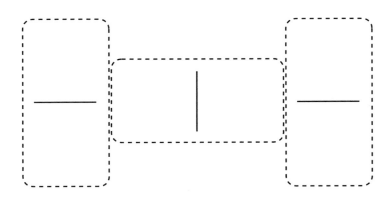

Surround

Task   Children identify the suit associated with a domino double.

Set-up   Children work independently using one set of dominoes and a "Surround" mat.

Start-up   Have children select any double. Then challenge them to surround the double with all the dominoes in the matching suit. Have them check to see that they have found them all.

Discussion   After children have identified the suit to which the double belongs, they will look for other members of the suit. Encourage them to enumerate the suit in order, from 0 to 6, to be sure that they have found all the members.

One possible solution:

Wrap-up   Key question for discussion or response in journals:

- What did you do to find all the dominoes you needed to surround a double?

Follow-up   Have children do the activity using a double-nine set, using nine dominoes to surround the double.

Surround

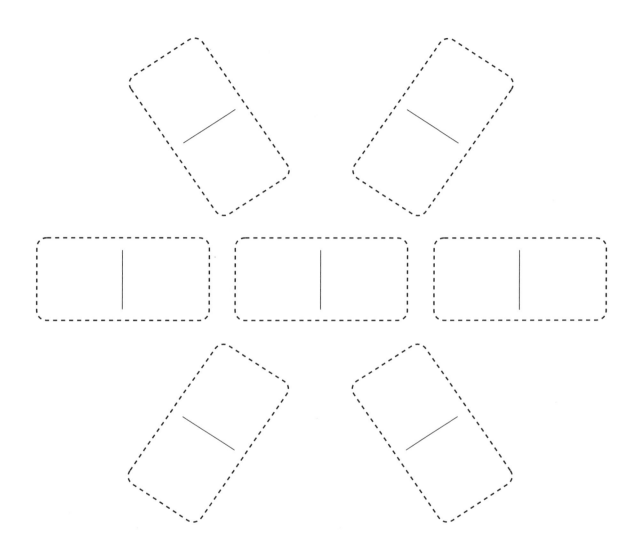

Find My Neighbors

Task

Children identify dominoes whose values are greater or less than the value of a given domino.

Set-up

Each pair of children uses a set of dominoes and a "Find My Neighbors" mat.

Start-up

Have children find the zero-suit dominoes. Then have them arrange the dominoes at the top of their mats according to their non-zero faces, from least to greatest. Tell children how to move one of the zero-suit dominoes into position on the middle of the mat. Challenge them to find this domino's "neighbors." Have partners take turns picking a domino from those displayed, then finding its neighbors. Children should repeat this procedure several times.

Discussion

Children will notice that the 0–0 and the 0–6 dominoes each have only one neighbor. Each of the other dominoes has neighbors with values of one more or one less.

Keep-up

Have children repeat the activity with each of the other suits. You may then have children randomly choose seven dominoes from a set and order them from least to greatest, according to the sum of their faces. Have them pick two of these dominoes, then identify the one or more dominoes that fall between them.

Wrap-up

Key question for discussion or response in journals:

• How can you tell if two dominoes are the neighbors of another domino?

Follow-up

Extend the activity by having children find a domino that fits between two neighbors. Have children work in pairs, taking turns selecting two dominoes and identifying those which fall between them. Encourage children to decide if just one domino or more than one domino fits between the two they have chosen. Ask questions such as, "How do you know which dominoes are between any two zero-suit dominoes? When can there be no dominoes between the two that were chosen?"

Find My Neighbors

Pick a domino.
Put it here.

↓

Follow the Leader

Task   Children compare the values of several dominoes to a single domino.

Set-up   A small group of children uses one set of dominoes and a "Follow the Leader" mat.

Start-up   Dominoes are placed facedown in the playing area. One domino is turned over, identified as the "leader domino," and placed on the mat. Each player picks a domino and turns it faceup. Children decide whether each of the dominoes picked has a value (the sum of its pips) that is less than, greater than, or equal to the value of the leader. One by one, these dominoes are placed in the corresponding columns on the "Follow the Leader" mat.

Discussion   Children will notice that each domino has a specific location depending on how it compares to the leader domino. Suggest that children then pick a new leader domino, leaving the other dominoes on the mat. With the new leader in place on the mat, have children rearrange the dominoes in the columns below to show how they compare to the new leader.

Keep-up   Have children record the relative positions of another group of dominoes for different leader dominoes.

Wrap-up   Key questions for discussion or response in journals:

- How did you decide on the column in which to place a domino?
- Why might the position of a domino change when a different leader domino is picked? Why might it not change?
- Which dominoes will never change positions even if the leader domino changes? Why?

Follow the Leader

LEADER

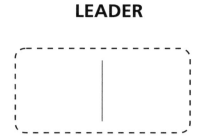

LESS THAN	EQUAL TO	GREATER THAN

Check It Out

Task   Children compare the values of three dominoes.

Set-up   A small group of children uses one set of dominoes.

Start-up   The dominoes are turned facedown in the center of the playing area. The first player picks three dominoes without looking and places them, still facedown, in a row. The two end dominoes are turned over and the values (the sum of its pips) are compared. The middle domino is then turned over. The players "check out" the sum of the pips on the middle domino to see if its value is equal to the values of the end dominoes or if it falls between them. If its value is either equal to or between the two, and the player correctly identifies it, the player wins the three dominoes. If the value of the middle domino is not between the values of the end dominoes, the three dominoes are returned to the pile. Play continues until all the dominoes have been won.

Discussion   The children will notice that the greater the difference between the end dominoes, the more likely it is that there will be one or more dominoes that can be positioned between them.

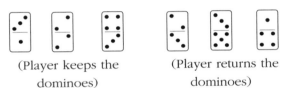

(Player keeps the   (Player returns the
dominoes)   dominoes)

Wrap-up   Key question for discussion or response in journals:

- If you could choose the end dominoes, which would you choose to be sure that there will be a domino whose value falls between them?

Sort and Classify

Children use observation, description, and classification as they sort dominoes in different ways. They represent their schema in graphs, dichotomous sets, and Venn diagrams and develop communication skills as they share their ideas.

Let's Graph!

Task   Children sort a set of dominoes according to the sum of the domino faces, thereby forming a graph.

Set-up   Each pair of children uses one set of dominoes and a "Let's Graph!" mat. Each pair will need scissors and tape

Start-up   Have each pair cut their mat in half along the dotted line. Tell them to tape the two halves together to form one continuous mat on which to place dominoes according to their values, from 0 to 12. Ask children to predict which value will be represented by the greatest number of dominoes. Then allow them to sort their dominoes by putting each in the column that names the sum of its faces. Tell children that they will have to position some dominoes off the mat as they build their graphs.

Discussion   Children will notice that the number of dominoes varies with the different sums. Some may observe that there is only a single domino associated with the values 0, 1, 11, and 12; that sums in the middle of the set have more dominoes associated with them than do those at the ends; and that there are four dominoes which have a value of 6.

Children's graphs will look like this:

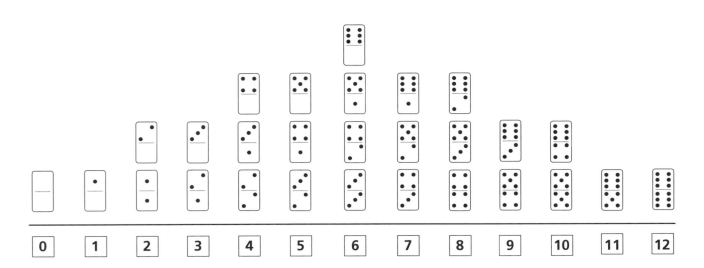

Wrap-up   Key questions for discussion or response in journals:

- Did you correctly predict which sum would have the greatest number of dominoes? Why or why not?
- Why are there more dominoes for some sums than others?

Let's Graph!

0	1	2	3	4	5	6

7	8	9	10	11	12

This Way or That?

Task   Children classify any six dominoes using one attribute.

Set-up   Each pair of children uses one set of dominoes and a "This Way or That?" mat.

Start-up   Have pairs choose any six dominoes from the set and place them in the "hopper" at the top of the mat. Tell children they will classify the dominoes according to their sums. Have children select two cards— "Has an even-number sum" and "Has an odd-number sum"—from the set cards on page 44; then place them on the mat. Tell children to slide the dominoes in the hopper down the appropriate chute. Repeat the activity with a different set of six dominoes.

Discussion   As they locate the dominoes according to their odd or even sums, children will notice that there are more even than odd sums. This is true because all pairs of even faces and all pairs of odd faces yield even sums, whereas only those pairs made up of an odd and an even face yield an odd sum. A sample game is shown.

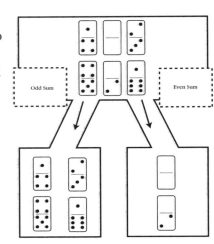

Wrap-up   Key questions for discussion or response in journals:

- How are all doubles different from all singles?
- Why are there more even-numbered sums than odd-numbered sums?

Follow-up   Give children others ways to sort or let them choose their own. Here are some possible pairs of attributes that you might use:

      singles–doubles

      sum of 5 or greater–sum of less than 5

      at least one face blank–no faces blank

This Way or That?

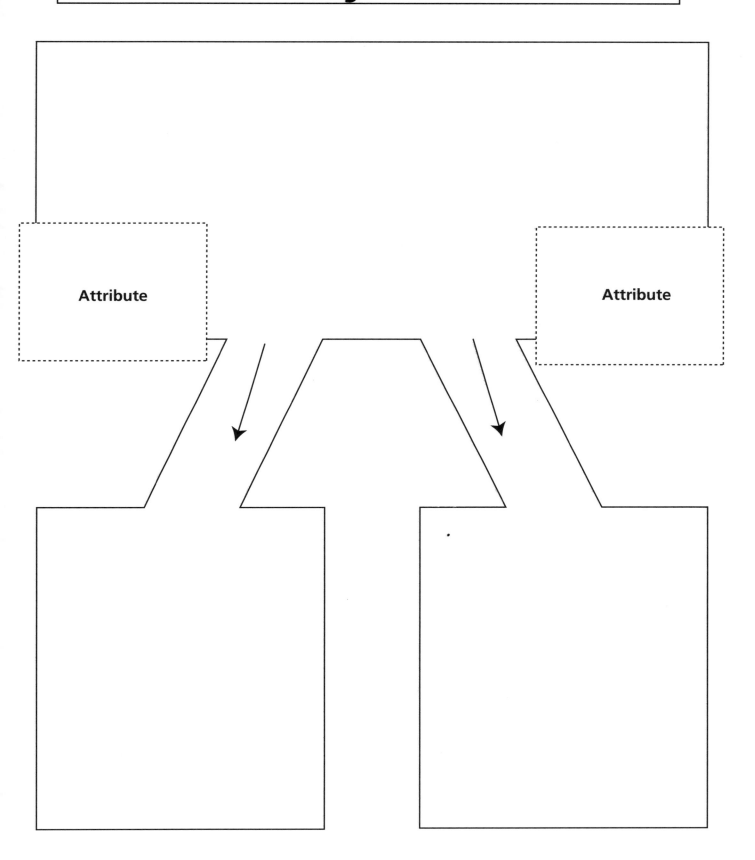

Attribute

Attribute

Math Marvel

Task   Children sort a set of dominoes according to their values and solve a simple probability problem.

Set-up   Each pair of children uses one set of dominoes and a "Math Marvel" mat. Each pair will need a paper bag.

Start-up   Tell children to imagine that they are going to play a game invented by a math wiz named Math Marvel. Read the game rules aloud:

Put a set of dominoes into a bag. Pick a domino from the bag and put it on the mat. Write the sum of its faces. If the sum of the faces is 4, 5, 6, 7, or 8, Math Marvel wins. If the sum is any other number, you win. Circle the winner.

Have children predict what the outcome of some picks might be, then have them find some dominoes that would score for Math Marvel and some that would score for them.

Play a demonstration game. Have children play the game for a total of five times.

Discussion Have children talk about their results and discuss whether or not the game is "fair." Make a class chart, combining the results of all pairs' picks. Then have children sort the dominoes into those with which Math Marvel would win and those with which they would win. They will find that Math Marvel can win with 16 dominoes and that they can win with 12.

Wrap-up   Key questions for discussion or response in journals:

• How did you decide on your prediction? Was your prediction correct?

• Are the rules of the game fair? Why?

Follow-up   Challenge children to change the game rules to make the chances of winning equally likely, or to make them "fair."

Sort and Classify

Math Marvel

Math Marvel wins with a sum of 4, 5, 6, 7, or 8.
You win with any other sum.

PICK	DOMINO	SUM	CIRCLE WINNER	
1ST			Math Marvel	You
2ND			Math Marvel	You
3RD			Math Marvel	You
4TH			Math Marvel	You
5TH			Math Marvel	You

Who Am I?

Task   Children use visual skills to identify a domino described in a riddle.

Set-up   Each pair of children uses one set of dominoes and a "Who Am I?" mat. Each pair will need a 5 x 8 index card.

Start-up   You may want to model the activity by showing the children any three dominoes and giving them two clues to help them identify one of the dominoes. Then draw their attention to the dominoes at the top of the mat. As you read the clues aloud, explain that each clue should help children eliminate one of the dominoes. Explain that the one that is left is the answer to the riddle.

Discussion Ask children how they solved the riddle at the top of the mat. Ask why the domino with faces of 4 and 1 was not the solution. Then ask why the domino with faces of 5 and 2 was not the solution. Finally, discuss why the remaining domino fits both clues—it has a face with 5 pips and a face with 4. Repeat this line of questioning for the set of clues that each pair makes up for the three dominoes they choose.

Keep-up   Children's riddles may relate numbers of pips or they may tell something about the numbers such as "The top face is even." Children exchange cards, then follow each other's clues to solve the riddles.

Two possible riddles:

1) Each of my faces is greater than 2. Each of them is odd. The sum of my faces is 8. Who am I? (Answer: 3–5)

2) The difference between my faces is 4. Both of my faces are even. Neither face is blank. Who am I? (Answer: 6–2)

Wrap-up   Key question for discussion or response in journals:

- How did you solve each riddle?
- Was it harder to write a riddle or solve a riddle?

Follow-up   Children each choose five dominoes and tell or write clues to describe one of them. Have them exchange their clues with their partners who try to solve their riddles. Key vocabulary may be discussed with children prior to writing clues.

Who Am I?

One of my faces has 5 pips.
My other face has 4 pips.
Who am I?

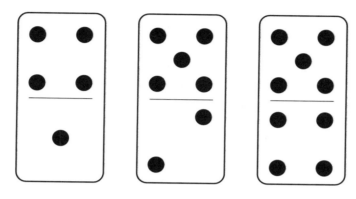

Choose three dominoes that are alike in some ways. Put them here.

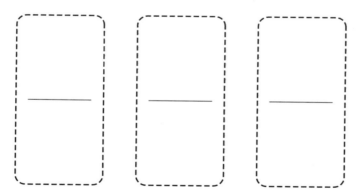

Write a "Who am I?" riddle for your dominoes.
Write your clues on a card.

Double-Circle Sets

Task   Children sort dominoes using a Venn diagram.

Set-up   A small group of children uses one set of dominoes, a "Double-Circle Sets" mat, and "Double-Circle Set Cards" (page 44).

Start-up   Play a demonstration game. Have children cut apart their set cards. Then, have them choose any two cards and put them on the mat. Have each child pick 6 dominoes. Players should take turns placing one of their dominoes into one circle or the other or into the intersection of the circles. After each domino is played, another is picked from the pile, until none are left. A picked domino that does not belong to either set is put down outside the playing area.

Discussion   Stress that the dominoes that meet the description of both set cards must be placed in the intersection of the circles. Point out that there will be times when no dominoes belong in the intersection. One sample solution is shown.

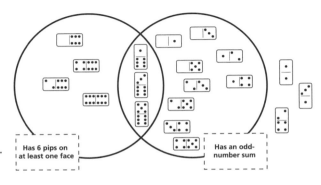

Has 6 pips on at least one face

Has an odd-number sum

Keep-up   Children may receive points for the dominoes they identify as members of the sets—1 point for each domino placed inside a circle and 2 points for each placed in the intersection. No points are scored for dominoes picked and put down outside the circles.

Allow children to write their own set descriptions on the blank set cards provided.

Wrap-up   Key questions for discussion or response in journals:

• How did you decide where to place a domino?

• What would the game mat look like if it was played with the set cards "Doubles" and "Has an even-number sum"?

Follow-up   Have children play "Triple-Circle Sets" with three intersecting circles. They may use the same scoring process except for scoring 3 points for each domino placed in the intersection of all three circles.

Double-Circle Sets

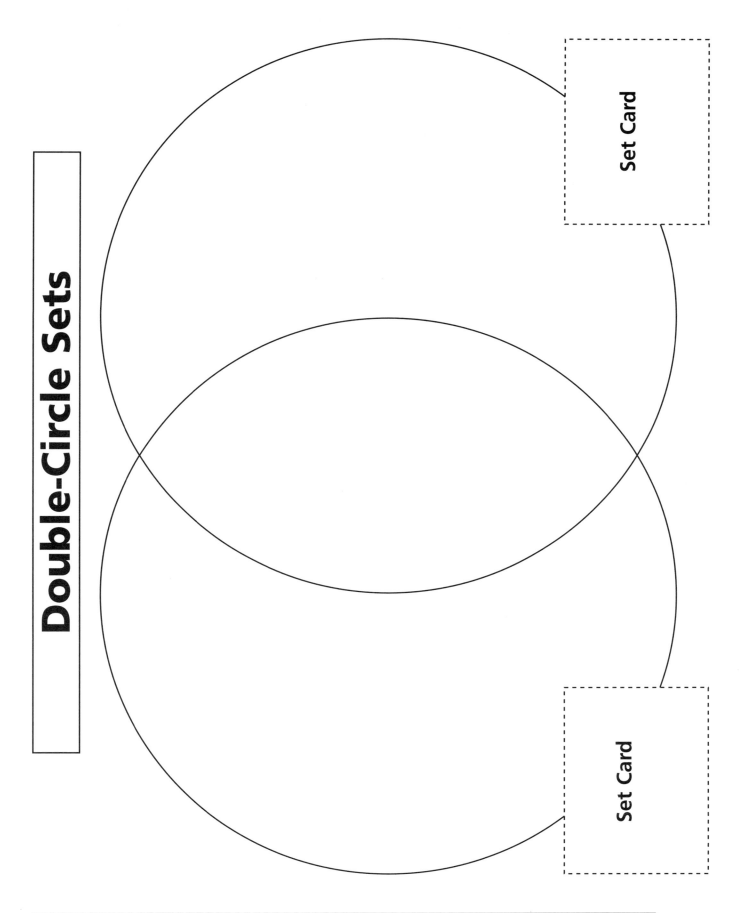

Set Card

Set Card

Double-Circle Set Cards

Doubles	Has at least one face blank
Has a sum of 7	Has an even-number sum
Has an odd-number sum	Faces have a difference of 2
Has 6 pips on at least one face	Has 4 pips on at least one face

Sort and Classify

What's It Worth?

Children combine and partition sets of dominoes as they solve problems and play games of chance. They engage in comparison and mental computation as they determine the relative values of dominoes. Children continue to develop the language of dominoes.

Is It a Tie?

Task   Children play a game to compare the values of dominoes.

Set-up   A small group of children uses one set of dominoes.

Start-up   Each child draws a domino from the set turned facedown on the playing area and determines its value by finding the sum of its faces. Children then compare dominoes to see whose has the greatest value. The child with the greatest-value domino takes all the others drawn. If any players have drawn dominoes with the same value, to break the tie they each draw three more dominoes, one at a time, as they together say "T-I-E." They compare the values of the dominoes drawn on "E." The player with the greatest-value domino takes the others drawn. Play continues until there are no dominoes left to draw. The winner is the player with the most dominoes.

Discussion   As they find the value of the dominoes, children will discover that some of the values can be obtained in several ways. They will also discover that other values can be obtained in just one way.

Keep-up   This game can also be played by finding differences. Have each child draw a domino and subtract the lesser face value from the greater face value. The child whose domino has the least value would take the others drawn.

Wrap-up   Key questions for discussion or response in journals:

• For each game, which dominoes will always win?

• For each game, which dominoes will always lose? Why?

Follow-up   Children may play "Greatest Sum," a game similar to "Is It a Tie?" Instead of drawing one domino and finding its value, each child draws two dominoes and finds their total value. The child whose dominoes have the greatest sum takes all the dominoes drawn. As children gain experience, the game may be played with each of them drawing three dominoes.

Play for Ten

Task Children play games combining dominoes in different ways to form sums of 10.

Set-up Each pair of children uses the 20 dominoes that remain when one each of the dominoes with values (sum of the faces) of 5 to 12 are removed from a set.

Start-up **1. Find Ten:** Eight dominoes are placed faceup in the center of the playing area. The rest of the dominoes are dealt out to the players. In turn, a player turns over one of his or her dominoes and tries to match it to a domino in the center to make a pair with a sum of 10. A player who makes a 10 takes the dominoes. If a turned-over domino cannot be used to make a pair, it is added to the center and the next player has a turn. The player with the most pairs wins.

2. Odd-Out Domino: To the 20-domino set, add the double-zero domino. Identify it as the "odd-out domino." Deal the 21 dominoes to all the players. Players should stand up their dominoes to form hands. Have them sort their dominoes, setting aside pairs of 10. Players take turns picking one domino from the hand of the player to their right, trying to make sums of 10. If no more pairs can be made, the dominoes are kept in the player's hand. Play continues until all possible 10s are made. The player who is left with the double-zero wins the game.

3. Fish for Dominoes: Seven dominoes from the 20-domino set are dealt to each player to form a hand of "fish." The remaining dominoes (fish) are placed facedown in the "pond." Players sort their hands, making pairs of 10 and setting the pairs aside. One player asks the others for a fish needed to make a pair with a value of 10. For example, a player with a fish whose value is 6 may say, "Do you have a 4?" If any player has a 4, a pair is made and the first player continues asking for fish. If not, a fish is "caught" from the pond and another player has a turn. The player with the most pairs of fish wins.

Keep-up All three games described above can be played with the complete double-six set to make pairs with values of 12. To play this form of Odd-Out Domino, remove the double-zero. Without the double-zero with which to make a pair, the double-six domino will become the odd-out domino.

Wrap-up Key question for discussion or response in journals:

• Was finding a pair easier for some dominoes than others? Explain.

Subtraction Action

Task   Children play a game in which they subtract the lesser number of pips on a domino face from the greater number of pips on the other face.

Set-up   Each pair of children uses one set of dominoes; a small marker, such as a paper clip; and a "Subtraction Action" mat.

Start-up   The marker is placed on the START space in the middle of the mat. One player selects a domino, determines which of the two faces has the greater value, and subtracts the lesser value from the greater one. After finding the difference, the player moves the marker that number of spaces toward GOAL ONE. The other player selects a domino, finds the difference between the faces, and moves the marker that number of spaces toward GOAL TWO. The player who reaches a goal first wins.

Discussion   Point out that when a double is picked, since the difference between two of the same numbers is 0, the marker cannot be moved. As the game progresses, children will notice that the marker will move back and forth across START until one player picks a series of dominoes with large differences between the faces.

Wrap-up   Key questions for discussion or response in journals:

- Which dominoes might you pick that would keep you from moving?
- Which dominoes is it best to pick? Why?

Follow-up   Players can select two dominoes, determine the total face value of each, and then move the number of spaces equal to the difference between them. Children can also design other game boards with different numbers of spaces and vary the playing rules.

Subtraction Action

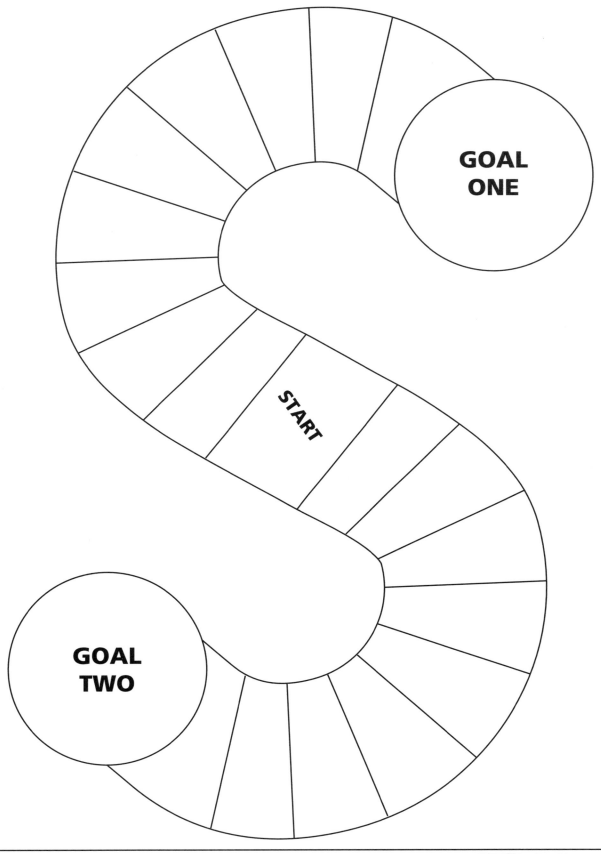

Math Activities with **DOMINOES** **49**

Diffy

Task Children play a game in which they find the differences between values of domino pairs.

Set-up A small group of children uses one set of dominoes and a "Diffy" mat.

Start-up Children select four dominoes and place them faceup on the "Diffy" mat. They find the difference between the face values of each pair of adjacent dominoes and write the difference in the circle between them. Then they find the difference between each pair of adjacent circles and write this in the square between them. Children then find the differences between adjacent squares and write them in the circles between them. They continue to find differences, as necessary, until all four adjacent numbers are the same.

Discussion Children learn that starting out with dominoes in certain arithmetic sequences or with dominoes having the same value will enable them to get equal differences in just one turn. However, selection of dominoes having other values may require as many as four or more turns to get equal differences in all positions.

One sample game:

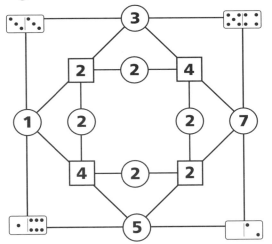

Wrap-up Key question for discussion or response in journal:

• How can you predict the number of turns it will take to get equal differences in all the innermost positions on the Diffy mat?

Diffy

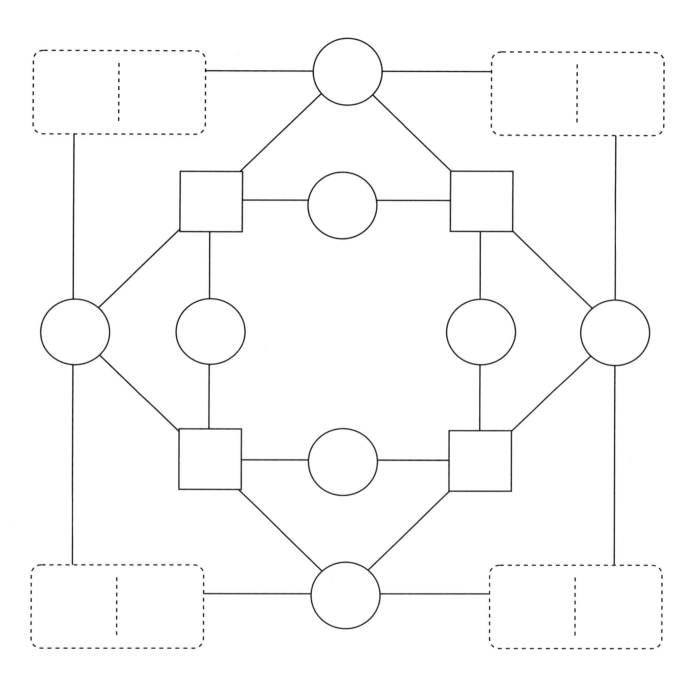

Pip-Pip-Hooray!

Task
In a variation of the game bingo, children associate the value of a domino with the total number of pips on its two faces.

Set-up
A small group of children uses one set of double-six dominoes and paper and pencil for recording number sentences. One child, the caller, has a set of 0 to 12 sum cards (page 80), one for each of the possible values of a domino. For double-nine dominoes, a set of 0 to 18 sum cards are needed.

Start-up
Children each select 4 dominoes, and place them faceup in front of them. The caller randomly chooses a number card and displays it. If a player has a domino whose value is equal to that number, he or she writes a number sentence for that domino, then turns it over. For example, if the number card for 7 is chosen, a player with the 5–2 domino would write 5 + 2 = 7, then display the domino. The first player to turn over all his or her dominoes calls "Pip-Pip-Hooray!" The number sentences are checked against the corresponding dominoes. If they are correct, that child wins.

Discussion
Children who do not have immediate recall of basic facts to 12 (or to 18) may need to compute the value of each domino before playing the game. The children should notice that some values will be called more than once because there are several dominoes with the same value. It should be decided before the start of the game whether more than one domino having the same value may be turned over for a given call. If this is allowed, additional time should be allowed between calls to find sums.

Wrap-up
Key question for discussion or response in journals:

• Which dominoes make it easiest to win? Why?

Follow-up
The game may be played with 9 dominoes arranged faceup in 3 rows and in 3 columns. The first player to turn over 3 dominoes in a row, column, or diagonal wins as long as all his or her number sentences are correct and match the turned-over dominoes.

Think It Through

Children use computational skills as they arrange multiple addends and subsets to solve problems. They engage in a variety of games with dominoes which encourages them to use mental computation strategies. Children develop creativity as they develop variations of the games.

Number Neighbors

Task   Children combine specific dominoes to form target sums.

Set-up   Children work independently or with a partner using one set of dominoes, a "Number Neighbors" mat, and a set of sum cards (page 80).

Start-up   Have children find all the dominoes with values of 3 and 4 and arrange them in this order: 0–3, 1–2, 1–3, 0–4, 2–2 on the mat. Ask children to find the sum of the faces for each domino and place the matching sum card below it. Identify all the dominoes as "number neighbors." Tell children to then write number sentences according to these rules:
- Use three or more number neighbors.
- Each number sentence must have a different sum.

Discussion   Children will be surprised to find that they can make many number sentences, all of which produce only six different sums: 10, 11, 12, 14, 15, and 18.

Wrap-up   Key questions for discussion or response in journals:
- How can you be sure that you cannot get a sum of 16 using these dominoes?
- If you change the order of the dominoes, will all the sums you found still be possible?

Follow-up   Children can repeat the activity using dominoes with values of 3 and 5, values of 2 and 4, and values of 1 and 6.

Number Neighbors

Put number neighbors dominoes here.

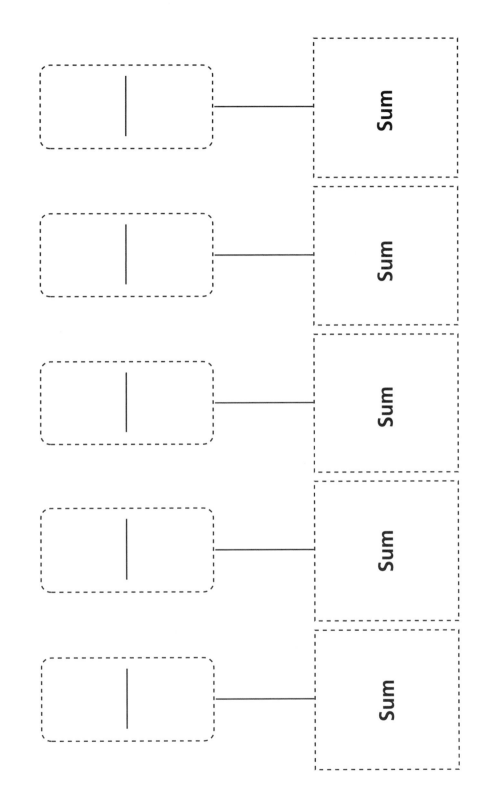

Match Mates

Task   Children play a memory game in which they identify pairs of dominoes with equal sums.

Set-up   A small group of children uses one set of dominoes and a "Match Mates" mat.

Start-up   As in traditional memory games, children find pairs of dominoes with equal values. They start by finding all the pairs, then mix them up and put them facedown on the mat. Children will find that eight dominoes cannot be used. Children take turns trying to turn up a pair of dominoes with equal values. If there is a match, the child keeps the dominoes and gets another turn. If there is no match, the dominoes are turned over again in the same location and another player takes a turn. The game continues until no dominoes remain in play. The winner is the player with the greatest number of pairs.

Discussion   To help children sort the set of dominoes for the mat, hold up the 0–0 domino. Ask children if there is another domino whose sum is also 0. Then hold up the 0–1 domino and ask if there is a domino with a matching sum. Continue this procedure, asking about the sums from 2 to 12, in order. Children will notice that for each of the values 4 through 8 there is an odd number of dominoes.

Follow-up   Children can repeat this activity by making pairs with sums of ten. Such a pair might be 1–3 and 2–4. Children will have to remove one each of the dominoes with values of 5 to 12 leaving a set of 20 dominoes for the game.

Other target sums are possible. If 12 is the target number, the whole set of dominoes will be used.

Wrap-up   Key question for discussion or response in journals:

- How were you able to remember where a turned-over domino was located?

Match Mates

Circle, Circle Sums

Task Children combine dominoes to form equal sums and represent them in a Venn diagram.

Set-up Each pair of children uses one set of dominoes and a "Circle, Circle Sums" mat.

Start-up Children place dominoes on the "Circle, Circle Sums" mat so that the sum of the values in one circle and the intersection equals the sum of the values in the other circle and the intersection. They try to find different solutions for all the possible sums.

Discussion You may want to model the activity using just a few dominoes, such as those with values from 0 to 6. After children have experience with these few dominoes, the whole set may be used. Children will discover that there are a variety of ways to make the various sums. At first, the children may use trial and error to find a solution. Later, they may use information obtained from one solution to arrive at a new solution.

Some sample solutions:

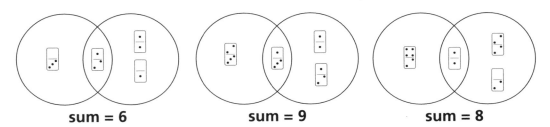

Wrap-up Key questions for discussion or response in journals:
- What patterns do you see as you try different dominoes?
- For which sums did you find the most solutions?

Follow-up More challenging formats may be used involving three intersecting circles and requiring that one domino be placed in each of the five sections thus formed. Students should have a recording sheet.

Some sample solutions:

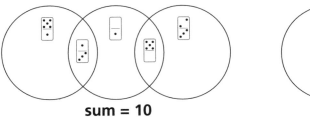 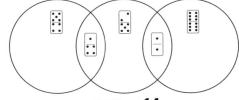

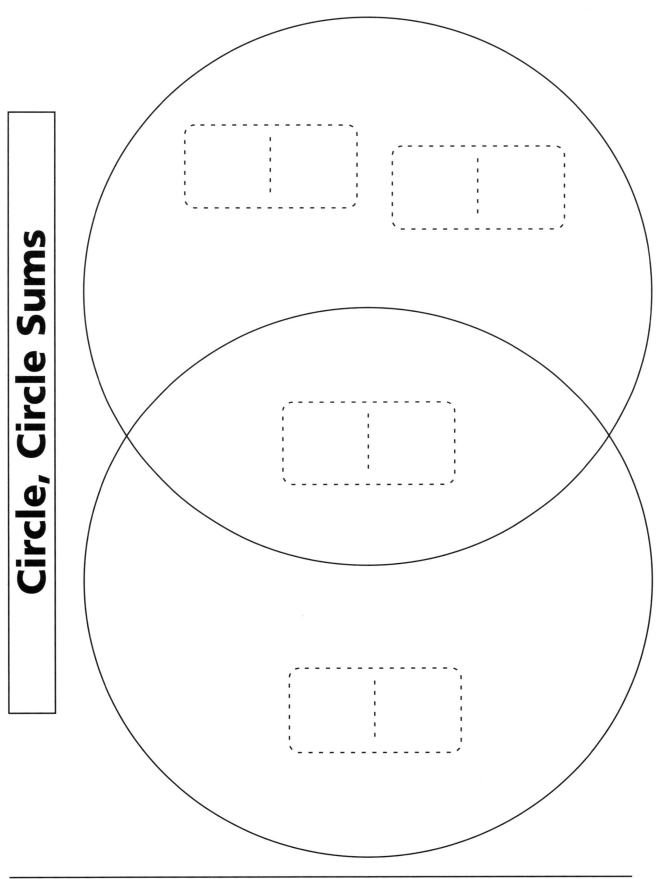

Circle, Circle Sums

Memory Mates

Task   Children play a memory game in which they find pairs of dominoes with equal sums or differences.

Set-up   A small group of children uses one set of dominoes.

Start-up   As in traditional memory games, children find pairs of dominoes with equal values. A match is made if the sum or difference of the pips on one domino is the same as the sum or difference of the pips on the other domino. Dominoes are mixed facedown in the playing area, arranged randomly in four rows of seven dominoes each. In turn, a child tries to turn up a pair of dominoes, each of whose faces have the same sum and/or difference. If there is a match, the child keeps the dominoes and takes another turn. If there is no match, the dominoes are returned to the location from which they were taken and another player takes a turn. The game continues either until no dominoes remain in play or if no more matches are possible. The winner is the player who collects the most pairs.

Discussion   To help children to "see" how the sums and differences are determined, display a domino and find the sum by adding its two faces. Then find the difference on the same domino by subtracting the lesser face from the greater face. It may help children to turn the domino so that the greater value appears first as in a written subtraction problem. Encourage children to practice with several dominoes, including doubles (which have different sums but always have a difference of 0) and dominoes with one blank, for which the sum and difference are always the same.

Point out, for example, that a 6–2 domino has a sum of 8 when the faces are added and a difference of 4 when the faces are subtracted. In a like manner, a 3–1 domino has a sum of 4 and a difference of 2. Therefore, a 6–2 domino can form a match with a 3–1 domino because they both can equal 4.

A MATCH

Wrap-up   Key questions for discussion or response in journals:

- How did you remember where a turned-over domino was located?

- What matches could you make for a double?

Target 20

Task   Children find different ways of arranging sets of dominoes to reach a target number.

Set-up   Each pair uses one set of dominoes.

Start-up   Children find dominoes, the sum of whose values equals 20. As they use trial and error, they make decisions about whether they need to replace some dominoes with those of greater or lesser values to make a sum of 20.

Discussion   Encourage children to form the least value possible with any two dominoes. Then tell them to substitute dominoes with greater values to increase the sum to 20. Some children will be able to start with any guess and substitute dominoes to get a greater or lesser sum.

Some sample solutions:

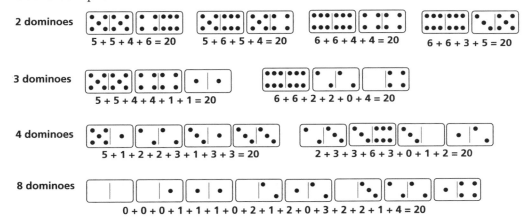

2 dominoes
$5 + 5 + 4 + 6 = 20$ $5 + 6 + 5 + 4 = 20$ $6 + 6 + 4 + 4 = 20$ $6 + 6 + 3 + 5 = 20$

3 dominoes
$5 + 5 + 4 + 4 + 1 + 1 = 20$ $6 + 6 + 2 + 2 + 0 + 4 = 20$

4 dominoes
$5 + 1 + 2 + 2 + 3 + 1 + 3 + 3 = 20$ $2 + 3 + 3 + 6 + 3 + 0 + 1 + 2 = 20$

8 dominoes
$0 + 0 + 0 + 1 + 1 + 1 + 0 + 2 + 1 + 2 + 0 + 3 + 2 + 2 + 1 + 4 = 20$

Wrap-up   Key question for discussion or response in journals:

• What helped you to get the sum of 20 using each number of dominoes?

Follow-up   Children may choose any target number between 5 and 42. Tell them to find a set of dominoes whose face values can be combined to obtain their target number. Then challenge them to find a different set of dominoes to represent their number in a different way.

Lucky Roll

Task   Children play a game in which they find the sum or difference of the faces of dominoes.

Set-up   Each pair of children uses one set of dominoes and one die or a 1–6 number cube.

Start-up   Each child picks 5 dominoes from the set, which is turned facedown. The first player rolls the die and gives up all the dominoes in his or her hand that have a sum or a difference equal to the number rolled. The second player uses these dominoes to begin a win pile. The first player then picks the same number of dominoes from the pile as were given up in order to maintain a hand of five. Then, the second player rolls and gives up dominoes. The game is over when no more sums or differences can be made. The winner is the player who has collected the most dominoes in his or her win pile.

Discussion   You may want to play a demonstration game with one child before having children play on their own. Some children may need to count the pips to find the sum and/or record the face numbers to find the difference.

One sample solution:

If a player rolls a 3 and has any of these dominoes, they must be given up.

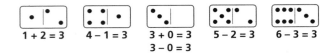

1 + 2 = 3   4 − 1 = 3   3 + 0 = 3   5 − 2 = 3   6 − 3 = 3
                          3 − 0 = 3

Wrap-up   Key questions for discussion or response in journals:

- What was a good number to roll? Why?
- Which dominoes will you never be able to put into your win pile?

Follow-up   To play the game with a double-nine set of dominoes, children will need to roll a pair of dice or two number cubes.

Connecting Paths

Children use matching, logic, and inference skills as they complete patterns to form domino paths. They make generalizations in order to form triangular and square paths. Children develop tolerance for a variety of solutions as they share strategies for completing paths.

Path of Nine

Task   Children match the faces of nine dominoes to form a continuous path.

Set-up   Each pair uses one set of dominoes and a "Path of Nine" mat.

Start-up   Have children look at the dominoes pictured on the mat and remove them from the set. Challenge them to use all nine of these dominoes to make a path by matching the faces end to end.

Discussion   Have children discuss how they decided which domino to put down next on a path when more than one domino could have been selected. If they get "stuck" as they make their paths, tell them to think about whether any dominoes can be interchanged.

One sample solution:

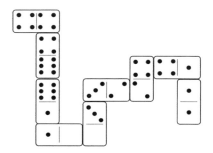

Keep-up   Have children pick nine dominoes, then try to make a path with them. Ask, "How many of the nine dominoes were you able to use to make a path?"

Wrap-up   Key question for discussion or response in journal:

- What did you have to do in order to use all of the dominoes to make a path?

Follow-up   Have children incorporate doubles into their paths by using a "crossing" to begin multiple paths. To make a crossing, children should place a double domino perpendicular to the rest of the dominoes in their path. Building on each face in the crossing creates two paths instead of one. Ask children, "How did a double lead you to change your path?"

Path of Nine

Find these dominoes.

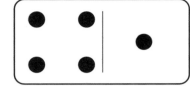 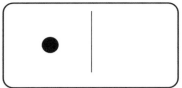

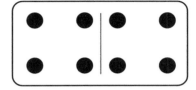 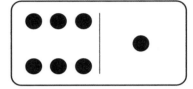 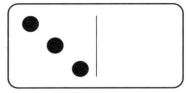

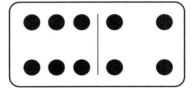 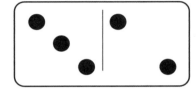 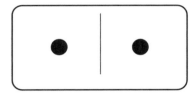

Use the dominoes to make a path here.

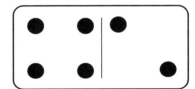

Hidden Path

Task   Children use logic to find dominoes to complete a path.

Set-up   Each pair of children uses one set of dominoes and a "Hidden Path" mat.

Start-up   Have children look at the dominoes shown on their mat and cover them with matching dominoes. Then have them find the dominoes missing from the path and place them on the mat to form a continuous path. Tell children that the blank dominoes on the path are not double-blanks, but represent other missing dominoes.

Discussion   As children complete the path, they will realize that where there is only one domino missing between two on the mat, the end faces of the two will tell them what the missing domino will be. Where more than one domino is missing between two on the mat, there is more than one possible solution. Children can record multiple solutions.

Keep-up   Some children may be able to create their own incomplete-path puzzles by first forming a path and then removing some of the dominoes along the path. They can challenge other children to find the missing dominoes from these newly created puzzles.

Wrap-up   Key question for discussion or response in journals:

• What clues did you use to help you finish the path?

Follow-up   Children who create their own puzzles may record them on paper, exchange papers with other children, then work each other's puzzles. Children's puzzles may also be used in a math center.

One sample solution:

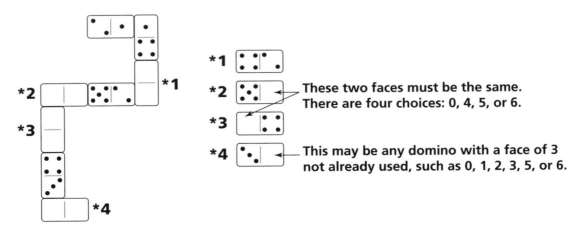

Hidden Path

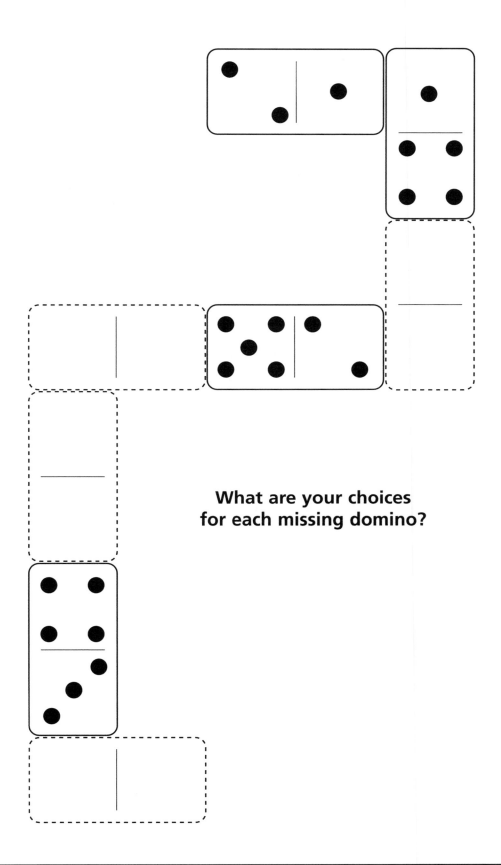

What are your choices
for each missing domino?

Pattern Paths

Task   Children use dominoes to identify and continue patterns.

Set-up   Each pair of children uses one set of dominoes and a "Pattern Paths" mat.

Start-up   Have children look at the dominoes shown on the mat and cover them with matching dominoes. Then have them find missing dominoes to continue each pattern. Tell children to compare patterns with their partners, then remove the dominoes, copying the faces on the mat to record the completed patterns.

Discussion Encourage children to look for a variety of patterns. Give them the opportunity to make their own pattern on the mat, remove one or two dominoes, then challenge other children to complete their pattern.

Keep-up   To help a child design a pattern, you may want to draw a few of the starting dominoes, then give it to him or her to complete.

Wrap-up   Key questions for discussion or response in journals:

- How did you decide which domino came next in a pattern?
- How did you plan a pattern of your own?

Follow-up   Explain that while the numerical order of a suit of dominoes may determine a pattern, a pattern may also reflect a double numerical sequence; for example: 6–1, 5–2, 4–3, 3–4, 2–5, 1–6. Point out that the way in which the dominoes are turned may also be considered in determining a pattern. Encourage children to explore patterns of various types.

Pattern Paths

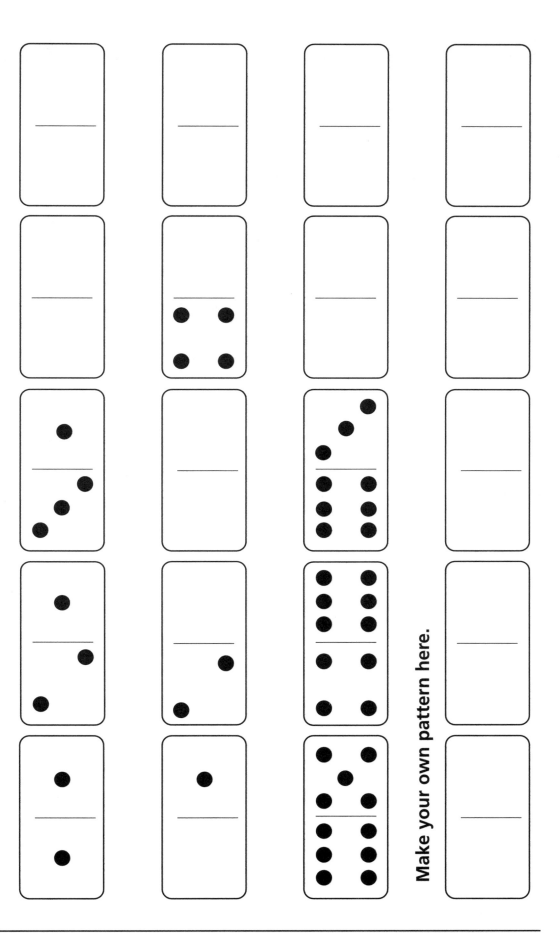

Make your own pattern here.

©1996 Cuisenaire Company of America, Inc.

Along the Path

Task   Children use dominoes to play a matching game.

Set-up   A small group of children uses one set of dominoes.

Start-up   Each child selects four dominoes to form a hand. One domino is taken from the pile and turned over as a starter. The first player tries to match one face of a domino from his or her hand to one face of the starter domino. If no match can be made, the player keeps picking dominoes until a match can be made. At this point, the next player has a turn. The game ends when one player runs out of dominoes or when no player can make a match. In this case, the player with the least number of pips left in his or her hand wins.

Discussion   Some children will notice that they may have a choice as to which domino to play. Encourage them to think ahead to their possible next play. Suggest that they consider how to force their opponent to pick dominoes from the pile by playing a domino with a face that they know their opponent does not have.

Wrap-up   Children who have completed the Follow-up activity for "Path of Nine" may now play "Along the Path" using doubles as "crossings."

Follow-up   Key question for discussion or response in journals:

• What are some things you can do to help you win the game?

Take Five

Task   Children match domino faces to form a continuous path.

Set-up   A small group of children uses one set of dominoes.

Start-up   Each child selects five dominoes without looking and tries to make as long a path as possible using only these dominoes. Any dominoes that cannot be used are exchanged, one at a time, for others from the pile. There may be as many as five exchanges. If no one has used all his or her dominoes after five rounds, the player with the longest path wins.

Discussion   Encourage children to make tally marks to keep track of the exchanges. Elicit the understanding that doubles are harder to play because each affords only one opportunity to make a match.

Keep-up   Children can be challenged to pick out five particular dominoes with which they can make a path.

Wrap-up   Key question for discussion or response in journals:

• Which dominoes are the most difficult to use in a path? Why?

Follow-up   More advanced players can continue the game for more than five exchanges, until all the dominoes have been played or until none of the remaining dominoes can be matched to any player's path.

Triangle Paths

Task Children match domino faces to form a continuous path in the shape of a triangle.

Set-up Each pair of children uses one set of dominoes. Each child has the top half of a "Triangle Paths" mat.

Start-up Children form a triangular path by matching two faces of two dominoes, then finding a third domino that matches the other two faces.

Discussion You may want to model the making of a triangular path. Match two dominoes that share a face, highlighting the common face. Direct children to find the third domino, the one that matches the other two faces. For example, if the 2–5 and 5–6 dominoes are matched at first, children will need to find the 2–6 domino to complete the triangular path. Children will notice that once two dominoes are played, the third domino is easily named. Some children will discover that it is impossible to use a double to form a three-domino triangular path.

Keep-up After children have had experience making triangular paths, each with three dominoes, they can try to make triangular paths with six dominoes each. Distribute the bottom half of the "Triangle Paths" mat on which they can make these paths.

Wrap-up Key questions for discussion or response in journals:

- Once you have matched the faces of two dominoes, how do you know which third domino will complete a triangular path?
- Why can't a double be used to make a triangular path with three dominoes?

Follow-up Children may want to explore making triangular paths using as many dominoes as possible.

Triangle Paths

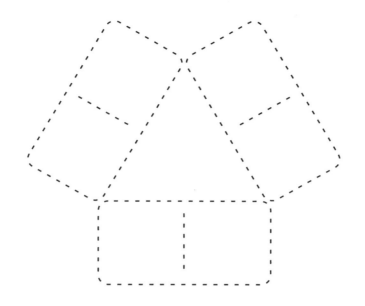

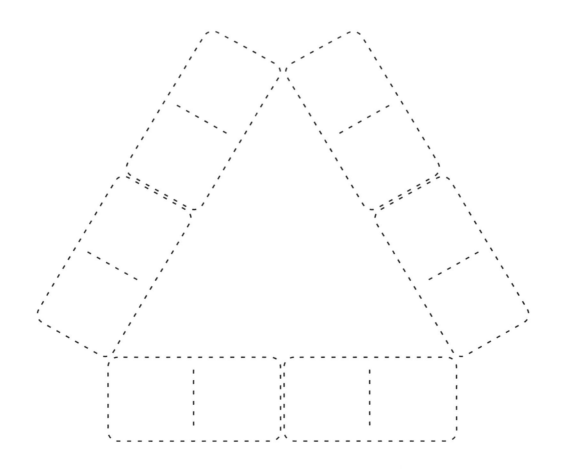

Math Activities with **DOMINOES**

Square Paths

Task   Children match domino faces to form a continuous path in the shape of a square.

Set-up   Each pair of children uses one set of dominoes. Each child has the top half of a "Square Paths" mat.

Start-up   Children use four dominoes to make a path that forms a square. They make as many of these paths as they can.

Discussion You may want to model the making of a square path. Match two dominoes that share a face, highlighting the common face. Direct children to find a third domino to match one of the end faces, then a fourth domino to match the two end faces. Children will discover that the missing dominoes are determined by matching faces of the surrounding dominoes. In forming their own square paths, children may find it easier to match corners first.

Keep-up   After children have had experience making square paths, each with four dominoes, they can try to make square paths with eight dominoes each. Distribute the bottom half of the "Square Paths" mat on which they can make these paths.

Wrap-up   Key questions for discussion or response in journals:

• How do you form a square path with four dominoes?
• Can doubles be used to form square paths? Explain.

Follow-up   Children may want to explore making square paths using as many dominoes as possible.

Square Paths

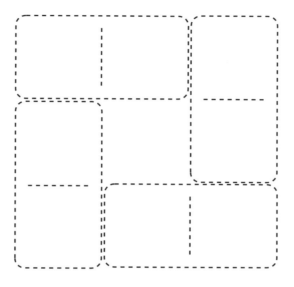

Equal Paths: Triangles & Squares

Task Children complete paths each of whose sides have the same value.

Set-up Each pair of children uses one set of dominoes. Each child has a copy of the bottom of the "Triangle Paths" mat and the bottom of the "Square Paths" mat.

Start-up Have children use six dominoes to make a triangular path. Then have them find the sum of each side of the triangle. If necessary, encourage them to exchange individual dominoes so that the sums of the pips on each of the three sides are equal.

Discussion Have children examine the dominoes that they first place on their mat. Have them decide if they need to exchange any dominoes with more or fewer pips. As they use trial and error, encourage children to build on the information they learn from each guess.

Keep-up Have children make several triangles with equal sides. Children may want to explore the range of triangular sums that it is possible to make, then decide which sums are impossible. Children can record their results as they investigate.

Wrap-up Key questions for discussion or response in journals:

- How did you go about making equal sums for the sides of a triangle?
- How do you know whether or not it is possible to make a triangle with sides of a given sum?

Follow-up Have children repeat the activity, this time using eight dominoes to make a square path with sides of equal value.

R.A.P. (Roll, Add, Place)

Task   Children subtract one addend from a given sum to find the missing addend.

Set-up   A small group of children uses one set of dominoes and a pair of dice.

Start-up   Children each take four dominoes from the set, which is turned facedown on the playing area. They place their dominoes faceup for their hand. One domino from the playing area is turned faceup as the starter. The first player rolls the dice, then tries to pair one face of a domino from his or her hand with one face of the starter, so that the sum of the faces equals the number rolled. If this is possible, the player puts his or her domino next to the starter and picks a domino from the playing area. If this is not possible, the player simply picks a domino from the playing area. The next player follows the same procedure. Play continues even when there are no dominoes left to pick. The winner is the player who runs out of dominoes first. If no one runs out, and no more matches can be made, the player with the least number of pips on his or her remaining dominoes wins.

Discussion   Some children may need to count up from the number of pips on the domino face to reach the sum. Other children may need help in finding the missing addend by subtracting the other addend from the number rolled. The more experienced child may realize that there is an advantage to playing greater-value dominoes toward the end of the game.

One sample game:

- The starter domino, 5–3, is turned faceup.
- The first player rolls a 4 and matches the 1–6 domino to the starter to show that the faces 3 and 1 have a sum of 4.

starter $6 + 4 = 10$ $3 + 1 = 4$

- The second player rolls a 10 and matches the 4–0 domino to the first player's domino to show that the faces 4 and 6 have a sum of 10.

Wrap-up   Key question for discussion or response in journals:

- How did you decide which domino to put down on your turn?

Spot Cards

(for 0–6 and 0–9 dominoes)

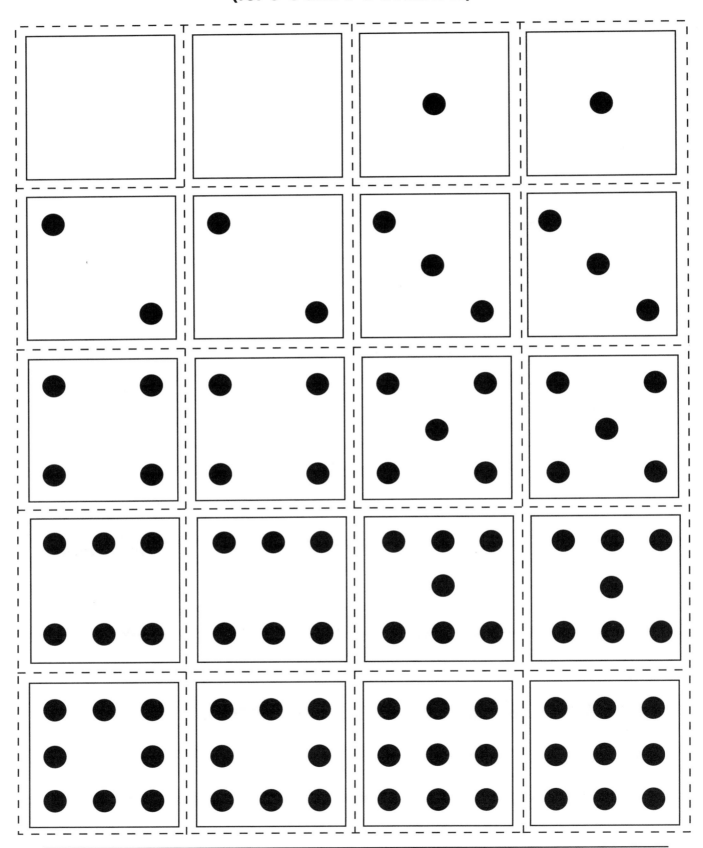

Number Cards

(for 0–6 and 0–9 dominoes)

0	1	2	3
4	5	6	7
8	9	10	11
12	13	14	15
16	17	18	

Sum Cards

(for 0–6 dominoes)

0	1	2	2	3
3	4	4	4	5
5	5	6	6	6
6	7	7	7	8
8	8	9	9	10
10	11	12		